THE REALITY OF GOD FORCES OTHER REALITIES

By
Elijah Muhammad
(Messenger of Allah)

Published by
Secretarius MEMPS Ministries
12685 Dorsett Rd #187 • Maryland Heights, MO 63043
Phone & Fax (314) 564-4003
Email: secmm@sbcglobal.net
www.memps.com

First Edition

ISBN 1-884855-46-6

Printed in the United States of America

DEDICATION

To all the Believers and Followers
of
Elijah Muhammad,
Messenger of Allah,
without whom, we as fractions
would have no expectation
of ever
becoming whole.

~ ~ ~ ~ ~

The
Messenger
is a Sword
In A Man's Hand

~ *Sayings of Elijah Muhammad, Vol. 3*

TABLE OF CONTENT

FREEDOM: IN THE ABSTRACT

"We must bear in mind that a definition is an instrument which helps us do our thinking and accomplish our purposes. A development is often frustrated merely because we are sticking to a former definition which is of no service to the new purpose.

Many subscribers to abstract definitions of freedom have settled with the belief that their government is nothing more than stewards of the people, and that such stewardship may be replaced at any moment. It is this possibility of replacement that gives the illusion of power - the power of appointing or voting. When the stewards position themselves in such a way as to acquire the power to control public opinion or "spin" the people to their will, then who's really in control? If the stewards stealthily redefine the definition of freedom, then would it be possible to make slavery as equally acceptable as freedom? Lastly, if this is possible, then it conversely stands to reason that freedom could be redefined to appear as slavery.

Is this is the real triumph of America? The political parties leading the masses, who, without an understanding of the meaning of words like freedom, become the blind might, and while believing they will have freedom through voting, non violence, civil rights and the like, infect the rest of the masses and other groups with this abstract idea; thereby rendering the majority of the so-called Negroes weakened once again and unable to grasp these ideas for constructive use and better service for their own people. The cycle goes on until the talk of freedom, justice and equality becomes constructively meaningless when it comes to being put in practical use for the advancement of the so-called Negroes.

"We have included in the constitution such rights as to the masses, appear fictitious and not actual rights. All these so-called 'Peoples Rights' can exist only in idea, an idea which can never be realized in practical life. What is it to the working class, bowed double over his heavy toil (his daily labor), crushed by his lot in life, if talkers get the right to babble, if journalists get the right to scribble any nonsense side by side with good stuff, once the working class has no other profit out of the constitution except only the pitiful crumbs which we fling to them from our table in return for their voting in favor of what we dictate, in favor of the men we place in power, the servants of our agentur...rights for a poor man are no more than a bitter piece of irony, for the necessity he is under of toiling almost all day gives him no present use of them, but on the other hand robs him of all guarantee of regular and certain earnings by making him dependent on strikes by his comrades or lock outs by his masters," as one writer wrote.

The so-called American Negro has accepted words for deeds; he is content with a show and fails to pause and note that the white man's promises are not followed by performance. Freedom in America is the right to do that which the law allows. It is not a sovereign right, but merely a granted privilege. It is simply a top-down political relationship; wherein, the Negro is not a sovereign of the land, and not even a citizen domiciled in any state of the union, but merely a piece of transferred property from the spoils of the Civil War. Since the law is in the hands of the sovereign white man, so one's freedom will thus be in his hands and his laws will abolish or create that which is desirable for the sovereign white man.

Gaining freedom, justice and equality, requires clear definitions that facilitate benefit for the black man is imperative.

"A development is often frustrated merely because we are sticking to a former definition which is of no service to the new purpose. We must bear in mind that a definition is an instrument which helps us do

our thinking and accomplish our purposes, "as before stated.

When we examine the importance of defining things, one may discover that the so-called Negroes in America has ingrained in his mind thoroughly, an education, words and the meaning of which prove ineffective for his own advancement; he is not able to put his thoughts into practical use, rather his thinking, if any, remains in theoretical routine.

America's counter sign against using ones thinking for his or her own people's advancement is force and make-believe.

"It is indispensable to undermine all faith, to tear out of the minds of the masses the very principle of Godhead and the spirit, and to put in its place arithmetical calculations and material needs."

It was the teaching of Christianity, the belief that God exists in three parts-the Father, the Son and the Holy Ghost, that became the black man's first education here in North America under servitude slavery. He was taught that God is a spirit and cannot be seen, only felt, and so being, God is a mystery-unknown. It is this way of teaching that enslaves the mind of the ignorant. This rearing on spooky beliefs, make-believe, ignorance of self and words that have no real benefit for the black man, is the root cause of his present condition.

The very words and definitions he has been given to do his thinking has enslaved him mentally. *"The method of slavery is so scientific, that even the English language is used to maintain it."*

It would be inadequate to define freedom, justice or equality without first defining What and Who God is, because the true meaning of these words have some interaction with the works of God.

"God means possessor of power and force," so teaches Elijah Muhammad, Messenger of Allah. This means that the one who is capable of exercising power and force over a thing in principle is the god over it. Yet, there is One Who is Supreme. With this definition, the actions of man can reach the level of God; in fact, God is man. The white man is well aware of this and performs like a god over the masses of the people. The fact that Man is God, which is the principle of Godhead that was torn out of the minds of the black man.

"All men are created equal" is an institutionalized phrase here in America; however, all men are not created equally. This is impossible to agree on and the white man knows this. There is superiority even in the animal world. There are animals who were created weaker than others. Some animals are vicious and some are harmless. Some are weak and others are strong. Man exists in the same way.

Equality means balance according to capacity. Each thing in this creation gets according to its capacity. Look at the California Cactus and the Redwood trees.

Both are plants; both plants require water and other nutrients for survival; however, a Redwood tree would die if only given the same amount of water as the Cactus need for survival; on the other hand, the Cactus would drown if given the same amount of water a redwood tree had to consume for survival. There is lesser and greater in the whole of the creation.

The so-called American Negro has pleaded to the U.S. Senate and Congress, the White House, and Courts of law for justice. He will continue to do this for generations until he discovers the reality of who God is and the knowledge of himself. Until then, his pleas will remain unanswered. Elijah Muhammad, the Messenger of Allah, by far and to date, is the only Blackman who has addressed this aspect in a concrete, calculable and practical way, but this would only be beneficial to one who has is willing to relinquish the white man's mind set of definitions.

Justice is the eventual working out of the will of God as indicated in the fundamental principles of truth. Justice is the antithesis of wrong, according to Elijah Muhammad. For each is that which it has earned and against each is that which it deserves.

It is new definitions and the use of them for a new purpose that will aid the so-called American Negro in a better future. The ability to distinguish between the truth and falsehood is worth more than gold today.

The Black man and woman needs a new mind, because the one they have has been conditioned in such a way that it doesn't serve their interests. Their minds have been conditioned to serve their former slave-masters. They can't think without him. They only think in ways that will always bring him benefit, even at the expense of their own. Their minds must be reprogram for the doing of self and helping self first before helping others. They have to be taught how to trust themselves with themselves and to trust each other.

THE UNSEEN CHAIN: WHEN SELF DISTRUST SELF

"Allah sets forth a parable: A man belonging to partners differing with one another, and a man (devoted) wholly to one man. Are the two alike in condition? Praise be to Allah! Nay, most of them know not."
(Holy Qur'an 39:29)

Have you ever had an instance when you said to yourself, "Did I see that," or "Did I hear you correctly?" or "Did I feel what I thought I felt?" Some of us have had to say this too many times to count; however, what's at play here is the act of second guessing one's own senses.

What's actually taking place when we second guess ourselves? We doubt or don't take serious the function of our senses. We take them for granted, because we have not built up a relationship of strict dependence on them. We question our senses, because there has been overwhelmingly numerous instances when we had to question them enough to not trust them fully. What instances could there have been, you might asks.

Take the television for example, how many times have you "seen" someone die and return from the grave, whether by magic or some other so-called supernatural method? The question is, "Did your eyes see them die or not?" Did you "see" Scottie, on Star Trek, beam Captain Kirk up or not? Did you see what you saw in the Matrix or not? Did your eyes see someone disappear or perhaps reappear? What am I getting at?

The eye doesn't possess the capacity to lie. It does not have a brain of its own. It's only a receptacle; an instrument which receives and sends information to the brain through the optic nerve. **Therefore, whether the stimulus is manufactured or real, the eye treats it the same. The eye**

doesn't distinguish between real and unreal; it only conveys the impression or experience to the brain, which in turn houses it and stores it for recall.

As a result of being conditioned on what's real and what's unreal (make-believe), we have grown accustomed to being chronic doubters. Our mental makeup is destabilized, because it is based on two or more opinions, two or more perspectives or varying, yet conflicting sources. Look at the educational system of America. It doesn't teach "truth," as such, it bases it conclusions on hypothesis, inferences, theories and case histories. Consequently, we have been taught to think the same way, which is why we rarely can agree on anything.

This dualistic thinking is a direct result of wrestling with reconciling the real and the unreal. The very fiber of our thinking is challenged from the time of infancy. It starts with ascribing language to animal; for example, talking birds, bears, and other animal within children's books and other form of media. The child then is nursed on television, which I will address more technically later, with doses of Sesame Street; wherein, they see larger then life "Big Bird" who talks, and Kermit the frog who talks, and a whole host of talking animals, who also have adults talking back to them. The mind is now being made to choose between what's real and what is not.

With more television come advanced forms of make-believe or unreality with horror movies, sci-fi, and various forms of things which are not real, but made to appear and sound real. By the time Christianity is introduced to them, wherein, they are taught that a so-called God, in the form of a SPIRIT comes from an **UNCONFIRMABLE PLACE** in the sky called heaven, then impregnates a PHYSICAL WOMAN, who is not his wife by the way, and begets a PHYSICAL BABY, then that baby grows up, dies and then comes back from death. After the before mentioned undermining of the child's mental fiber and capacity to distinguish between real and unreal, what fiber would they have left to refute the greatest make-believe story ever told? Their minds are saturated with uncertainty, even to the disbelief of their own senses; therefore, how could they resist such an onslaught of unreality without a firm grasp of what reality actually is?

It is this state of mind, which makes **doubting one's senses a "natural" mode of operating**. To operate in this way is to operate no differently than an animal of the field. Animals of field are only concerned with eating, procreating and some place to sleep. When one is rendered on the level of a mere beast, their major concern is food, satisfying their urges and a roof over their heads.

As mentioned earlier, **one tends to think that "truth" is that which occurs often, what is, or that which is widely accepted by the most people, or a regional consensus of the people (tradition)**. In the process of testing phenomena, that which results most often is most often considered "truth." This process has been taught and instituted within the Western World's concept of what's right and what's not. This world as we know it, makes up "truth" as they go, and as they feel it should be. In fact, they will bend anything to suite their interest, objectives and desires, even if its contradicting. Just because this method is widely used and enforced doesn't mean it's correct. Yet, being victims to this chain of events has made many give up and assume that there is no representation of anything contrary to this way of thinking; therefore, it must be all that there is. Nothing could be further from the truth.

If one were to ask you for a definition of "truth," what would come to your mind? Perhaps a few clichés' such as, "One man's truth is another man's lie, or truth, like beauty, is in the eye of the beholder, etc...." Yet, we have not dealt with the definition. In scripture, we are told that God is the Truth; in another place it is said that the universe was created in truth, and yet in another place it is told to us that if we know the truth, it will set us free. Let's examine what Elijah Muhammad, the Messenger of Allah (God), says about this aspect.

The Messenger teaches that God created Himself, and that all after that, came from Him, through Him, and designed by Him, meaning that after creating the very concept of creating with Himself being the first example or manifestation, everything else were simply products of Him or parts of himself. Therefore, if anything else exists other than what the Creator of this universe has created, then the burden of proof rests on it to manifest itself; otherwise, the God of Creation will be given all the credit.

In the Messenger's book, Our Saviour Has Arrived, he points out various illustrations of how the God of Creation created himself. **What's key here, however, is the fact that since there was no other god before Him, nor was there anything before Him, He then made everything the way He wanted it,**

designed everything the way He wanted it, and gave it place value accordingly.

The Arabic word to describe "Lord of the Worlds" is Rabb, which conveys not only the idea of fostering, bringing-up, or nourishing, but also that of regulating, completing and accomplishing; that is, of the evolution of things from the crudest state to that of the highest perfection. According to Islamic lexicologists, Rabb signifies the fostering of a thing in such a manner as to make it attain one condition after another until it reaches its goal of completion. Hence Rabb is the Author of all existence, Who has not only given to the whole creation its means of nourishment but has also beforehand ordained for each a sphere of capacity and within that sphere provided the means by which it continues to attain gradually to its goal of perfection. (M.M.Ali.)

Where man is concerned, he is considered the highest creature on the planet, because of his ability to think and reason. Thinking is the process of taking in various experiences through the senses and cross referencing them in combination of ways to function. Whereas, it is said in scripture that as a man thinketh, so is he. In another scripture, it is said that man is nothing but what he strives for, or what his own hands bring about.

Man is predicated on thinking. He is an animal aright, but thinking separates him from others by the process of being civilized. One having knowledge, wisdom, and understanding are just a few of the things which qualifies onc to be considered civilized.

The thought process is necessary to function; whereas, it is assumed that thought proceeds action; therefore, it could stand to reason that the active, physical world, had to be thought of or "conceived" first before it was created, because the unmatched degree of order, "Natural Order," strongly supports this. So much so, it sets the basis of mathematics.

Mathematics is a system of calculating, measuring, concluding or "adding-up." As mentioned before, with many experiences coming into ones mind, they must be stored, recalled, formulated, weighed, etc... to come to various conclusions in order to function or conduct one's self in various fashions. The conclusion to go left or right, up or down, towards or away, comes from a complex, super-fast process of weighing, remembering (reconnecting members) and decision making with the potential of coming up large quantities of ideas, and possibly action, there must be some standard of reference established or there would be complete confusion and chaos. This standard or universal reference point is "Nature" or the "Natural Order."

We refer to this standard or reference point as "reality." The Creator designed everything, gave it a place and value long before it reached a completion or perfection. All things start differently than it ends up. Therefore, we see that it must start with a thought. If one's thinking is damaged in such a way as to either deprive it of the capacity to use the natural order to aid in its thinking, or perhaps if one is merely kept ignorant of it usage, the logical and natural result is to be rendered on the level of common animal or beast of the field, who more than likely can be content with a little food, urge satisfaction and a place to sleep. Actually, this is considered fine living in many circles.

Animals are generally known for relying more on their instincts than intelligence. In the case of man, his emotions and feeling are stronger and serves as substitutes. Emotions are the weaker part of man. Once emotions dictates our conduct, rather than intelligence, we are living the level of beasts.

To be able to calculate, come to a conclusion, and have the confidence to act on the conclusion is the whole basis of being human. Is not this mathematics? Isn't it interesting that we can convert a word problem to a mathematical equation and an equation to a word problem? By submitting to this process, the result produces a condition of peace. Remember now, the Creator created us in such a way that by condition after condition, we evolve from crude states to states of completion or perfection; however, we must submit to the means and methods we were created out from; otherwise, we will have to perish from this state, create our own universe or system, recreate ourselves according to it laws and principles, then we can do as we please. Since this isn't so, we must abide and comply according to our composition, because form dictates function.

Without a grasp of what this system is all about, it is impossible to live according to it. Thought and product is the cornerstone of this creation. Nothing can be produced in this creation without there first being a thought. This one-to-one correspondence in this created reality, which we call the universe, is the definition of truth.

Let us not forget that to either not use this process or not know of its existence renders one on the level of beasts. For beasts and animals do not rely heavily on memory and recall. Their instincts suits them fine, but man is a different kind of animal. Once he or she stops using or never develops this reliance, they lose their right to called human.

Once he or she doubts their own senses, it shows that they now use another references of judgement, or other standards of reference. This ultimately relieves them of the very element necessary to keep them human. Once relieved of where the buck stops or a universal standard, the rule becomes what the majority says, or what the crowd does, or what the statistics point to, or merely how one feels. This is a sure recipe for confusion.

Elijah Muhammad, Messenger of Allah, demonstrated the process of not only recognizing a one-to-one correspondence, but masterfully and effectively acting on the results, which, in and of itself was a demonstration of the degree of his conviction he had about his thinking. This was demonstrated with his meeting with our Saviour, Master Fard Muhammad, and first recognizing Him, then confronting Him with surety. This combination of recognition and action definitely manifested calculation which made the Messenger conduct himself the way he did, which was to approach Master Fard Muhammad, Allah (God) in Person. The rest is history.

With thoughts from the first chapter in the Messenger's book, Message To The Blackman In America, wherein, he wrote, "For thousands of years, the people who did not have the knowledge of the person, or reality of God, worshipped their own ideas of God....The law of mathematics will not allow us to put three into one. Our nature rebels against a belief of God being a mystery and yet the Father of a son and a Holy Ghost without a wife or without being something in reality. We wonder how can the son be human, and the father a mystery (unknown), or a spirit?"

After completely undermining the thinking of the people with various forms of make-believe and untruths, then injecting them with Christianity or any other type of belief system wherein the God is mystery, will ultimately render any adherer on the level of a beast; for as the Messenger states, that adherence to the unknown is a setback for any people.

One may think that this article is perhaps an exaggeration; however, when one understands the method with which the truth or the light, as those into higher knowledge refer to it, has been hidden, this article won't be brushed off as a personal opinion so easily.

According to the Messenger, and his quoting from Edwin R. Embree's book, Brown Americans, he tells the story of law against education for the so-called Negroes. He quotes, from one of the Virginia

Delegates in 1832, "We have as far as possible, closed every avenue by which light may enter the slaves' mind. If we could extinguish the capacity to see the light, our work would be complete; they would then be on the level with the beast of the field and we should be safe. I am not certain that we would not do it, if we could find out the process and that on the plea of necessity."

The physical shackles are gone, but as the slave-master's children are more scientific, so are the chains.

Despite how many of us differ over the date which black people were brought to this country, facts state that they were tricked, kidnapped and brought here for no other reason than slave labor. This free labor, or as some called it, peculiar institutionalization, produced much wealth and international trade potential for the slave-masters, especially for those in the South. This perceived edge in wealth and potential political power by the brother in the south, facilitated a belief in the South that the South could do without the North. The great wealth produced from free slave-labor and the rich producing soil of the South, gave the South a commodity (food) that could be traded anywhere. With a spirit of independence running ramped over the globe, this made it difficult for the brother in the South not to think of succession from the Union; likewise, this made it difficult for the brother in the North to ignore such a potential.

This ballooning capability increased the fear in the North, which resulted in a civil war between North and South. With the "ploy of promises" to free the slaves, the Northern brother caused many southern slaves to abandon the South for a promise of freedom in the North.

Victory of the North brought about a signing of the Emancipation Proclamation. According to Booker T. Washington, "After the signing, there was lots of joy and jubilee back at the shacks and shanties, with great shows, and wild scenes of ecstasy. This lasted only a short while. The great responsibility of being free, of having charge of themselves, of having to think and plan for themselves and their children, seemed to take possession of them. It was like turning a youth, 8 or 10 years old out into the world."

"The great question with which all civilized nations of the earth must grapple with had now been thrown upon our mothers and fathers to be solved: "The question of: a home, a living, the rearing of children, education, citizenship, and the establishment and support of institutions."

"The elderly was hit the hardest, for their best years were gone. Gradually, one by one, stealthily (secretly, covertly) at first, the older slaves began to wander from the slave quarters back to the "big house" to have a whispered conversation with their former owners as to the future. There were many who tried independence and were able to squeeze out an existence for themselves."

This situation created two schools of thought that has transcended time and generations among the so-called Negroes, namely within the integrationist school and the separatist schools of thought.

Despite both schools of thought, whites' historical hatred of blacks have forced the slave-master and his children to set up unjust systems, ethnocentric cultural circumstances and hostile attitudes to ensure their "safety" against them.

Examples in Messenger Elijah Muhammad's book, Message To The Blackman In America, he cites Edwin R. Embree's book, "Brown American," that "The burning of Henry Lowry in Arkansas, proceeded by inches. Leaves soaked in gasoline were heaped about in small bundles so that torture would be dragged out. Ralph Roddy, a reporter, described the entire orgy in the Memphis press of January 27, 1921. He was able to cover the story because plans for the lynching had been made well in advance. The newspapers were notified to be ready to issue extras."

"When Henry Smith was burned at the stake in Texas, excursion trains were run for the event. Many women and children were in the throng that gloated over the suffering of the victim"

The above depiction is not necessarily intended to provoke the reader's emotion, but to illustrate a particular point: After thinking about what you and I have just read, let us not focus on the terror of the event per say, but on the mind set of those who were scheduling excursion trains, or those sending the news for the scheduling, or those running the extra editions of the newspaper, or those making sure they got front row seats, or those who were smiling while they posed for pictures with a backdrop or background of a completely burned and charred black man. If they had just one ounce of **NATURAL** human compassion, that element of compassion would not have allowed such one to stomach a heinous events of that caliber on such a common and consistent scale. Compassion is a natural emotion; therefore, if we reflect on the extreme degree of human atrocities that whites have committed, as sport, fun and pass-time, not just on darker peoples alone, but even on themselves, they therefore must be, as a whole, missing this natural human element, for their history verifies this fact.

How many of us could stand to see a dog, or any type of animal burned alive, watching as it moans and groans from the torment and agony of being heinously destroyed. If this is how you and I may feel about a dog or the like, what about a fellow human being - oh, did you say human being? Here exactly is my point, we, in white folks' mind weren't human. This is the type of mind-set that would allow them to "rationalize" such hideous actions, and to date, still perpetuate them.

Let it be understood that this frame of mind wasn't limited to adults only, for not only were there adults taking picture and gloating over the victim, but their children were in the throng and were very much so being groomed for the future; for instance, Edwin Embree, also on the same page, mentions what Walter White, deceased secretary of the NAACP, said he heard and saw in Florida, in his book, "Rope and Faggot." White recounts the gruesome tale of lynching in this country. While investigating an atrocious riot in Florida, White was met, he says, by three clean healthy children (white) headed for school. None were over 9 years of age. They gleefully described the event and "the fun we had burning the nigger."

If these children were 9 years old in the 1920's and gleefully talking about "the fun we had burning the niggers," what were they saying by the 1950's at approximately 40 years of age? Even in the 1920's, such norms were not new. There were others perpetuated from times far before that; "Read some of their books as to what was done against our slave parents by the fathers of these modern murderers of ours. As mentioned above right in the legislative bodies of the government: "We have as far as possible, closed every avenue by which light may enter the slaves' mind...If we could extinguish the capacity to see the light, our work would be complete; they would then be on a level with the beast of the field and we should be safe. I am not certain that we would not do it, if we could find out the process and that on the plea of necessity."

Now, if the children were personifying this spirit, a spirit well cultured and transferred at least from 1832 till 1920, well over 88 years, how much more would such a spirit be potent and powerful here in 2007, which is less than 87 years later than that - not to mention what it was like in the 50's or the 60's?

The institutionalization of such race hatred has made it more advanced and sophisticated and hard to detect from an outward or overt showing. Its perpetuation has been laced in various forms of media. I'm speaking more of the complexes of both peoples: The superiority complex of whites and lighter people of color, and the inferiority complex of darker peoples.

There has been many scientific methods used to maintain and nourish both complexes unconsciously; for example; take the menstrual show; it was made well known by a whiteman by the name of Al Jolson who would paint his whole face black and paint his lips a wide white color. This would amuse white audiences, but simultaneously and subconsciously, it would be fortifying and institutionalizing whites' hated, loathing and despise for what Mr. Jolson was mocking - you and I. It didn't stop there; further degrading and debasing would systematically take place to keep you and I believing we were not worth anything. To help drive this point home, many of our earlier actors would perpetuate roles to concretize these degrading images in our minds.

Roles that portrayed us as Ain't Jemima, a black, bandanna-wearing, big, fat manly mama, who runs master's kitchen and nurses all his babies; then you had a shiftless good-for-nothing Coon; then there was a happy without-a-care-(or responsibility)-in-the-world Sambo; then there was slick and sly, hambone. We can never forget Uncle Tom who would rather give his life to protect his master's life, then there was the young kids who were considered pick ninnies; and lastly there was the character of the stupid buffoon, like the character of J.J. in "All In The Family," Urcle in "Family Matters" and the ever abundant meal-ticketer parading before our eyes today).

Many of the programs on T.V. today perpetuate these images which fortifies the mind set on both sides without many of us ever knowing it; however, those engaged into the control of mass opinion making (advertisers, movie producers, etc...) don't even make programs without consulting this school of thought.

Although we were always hated, we still made strong, free labor; yet, as the pressure from slavery mounted upon our fathers and mothers, just out of sheer necessity and self-preservation, many of us invented machines and methods to keep the white man from killing us outright, but little did we know that he would steal all he could and eventually we would invent ourselves out of being useful.

Now that all the railroad tracks are laid and the swamps are drained and the whiteman and his woman has stolen enough song and dance routines and have put our natural musical prowess to digital synthesizers, he doesn't need his niggers any more, and he is forever moving towards making the so-called Negroes go away. You see, when Elijah Muhammad stated that one day the devil will drop you, he meant it. He will not be made into a liar!

We don't know that we can't see, because our definition of light and sight is faulty. We don't know that we don't know.

With the whiteman having complete control and access to books and any reading materials, especially scripture, and not allowing us to even learn how to read, as humans, this deprivation rendered us spiritually and mentally blind to the extent that not only did we not know that we didn't know God, or the white man (devil), but we didn't and still don't know that we are ignorant to the knowledge of ourselves. Consequently, our extreme craving to be accepted (which is natural since we are social creatures) and to be somebody, forced us to take on any identity we fancied or are given the opportunity to assimilate.

The fact that we have been wrestling with different names for some time is an apparent indictment that we just don't know who we are. Such a state can only be perpetuated if we don't know God (The Giver of Orientation). And a people who hasn't had a spiritual intercourse with God is considered spiritually virgin and mentally dead.

REALITY

"As we near the exhaustion of the Wisdom of this world which has not been able to shed enough light on our path in search for that Supreme Wisdom to keep us from stumbling and falling, we now seek the wisdom of Allah, The Best Knower and Guide in the Person of Master Fard Muhammad (to Whom be praised forever)...."

The above words of the Honorable Elijah Muhammad, Messenger of Allah, are very befitting, because they state that which many of us already know and that is, we are in the darkness about a great many things and now need the right knowledge to cut through the darkness. We, as well as our great grand parents have tried everything, but have come up empty handed; therefore, now that what we have been trying have failed and is still presently failing, it is a wise decision to consider the path of One who is the Best Knower.

As a result of calculated mental destruction and death, many believe in some type of formless spirit, to which we ascribe the name god. Whether we are Orthodox Muslims, a part of Christendom, Hinduism, Buddhism, etc..., the basis of whatever we perceive as the "god" or "all that there is" or "the great grand father or "divine mind" if you will, is without shape or form in our minds. This would make any sober minded person eventually ask: Who is this mysterious god or force or spirit that I have allowed myself to believe is controlling me and my actions? There is a form of control in all of our lives - those of us who are "normal," but does it come from an outside formless source?

Surely there are those who say there is no god; however, they have as yet identified that which they say don't exist; therefore, is such a conclusion valid? Simply, we know that there is some form of control in their lives, because if they didn't obey someone: parents, government, etc..., they probably wouldn't exist long in a civilized society.

The point here is that most civilized people operate according to some type of control and accountability, and we need to know if it comes from some force, spirit, or god we can't account for?

When dealing with what controls us, our actions and aspects of right and wrong surface. Doing right or wrong are ethical issues, which have followed all civilized societies; yet, before such issues can be addressed, there must be some established reference point to make a judgement. In other words, one needs to know where the buck stops or what or who has the final word.

In the society we presently live, most Black people suffer from what's known as a hysteresis: The failure of a thing that has been altered by an external agent to return to its original state when the cause of the alteration is removed. In other words, the failure of the black man and woman, who were made slaves, to return to their original state once the condition is removed. The white race (external force) altered black people from aboriginals into a people who do not even know that they don't know themselves. They were experimented on, altered and rendered on the level of beasts, permanent servants and today's free-slaves for white folks as a result of chattel slavery.

Putting them back in the right state is a tremendously hard task. It is not as easy as it sounds for them to be raised up from this dead state, because the mind is enslaved, which keeps the subject thinking like one. The method of slavery is so scientific, that even the English language itself is being used to maintain it.

Did you know that we are often deprived of very useful thoughts merely because the words which might express them are being temporarily preempted by other meanings (subordinated)? Or that a development is often frustrated merely because we are sticking to a former definition which is of no service to the new purpose? We must bear in mind that a definition is an instrument, which helps us do our thinking and accomplish our purposes.

When we examine what thought is, we simply come to the understanding that thought is the result of an impression made on the brain or in the mind. This is done through the senses: seeing, touching, hearing, smelling, and tasting. Each one of these, carry signals to the brain and the brain interprets the given experience and logs it as if on a shelf for future recall. When these senses are reaching out and relating with its environment, the question then becomes, is the environment on one part of the planet earth any different, in principle, than another environment on another part of the planet? No. For the basic elements are consistent around the earth, especially when it comes to how we as humans relate to it. It is so consistent, that we can carry our environments with us because the laws and principles governing this planet are just as consistent as the materials it governs. This whole scenario is called reality, which not only takes into account this planet, but all planets and life in this universe, which brings us back to our title: The reality of God forces other realities: New Thoughts and New Actions.

It would be extremely inadequate and an outright waste of time and resources, on the Creator's part, to create man with the power of thought, then make it an absolute necessity for the man to not only know Him, but to obey Him for longevity of life, then turn around and make the means by which we could know him - through our senses - ineffective and incapable of ascertaining Him.

In other words, we are told all our lives that if we don't serve God, we will be destroyed; yet, it appears that the only means and methods which the Creator gave us to KNOW anything are ways aside and completely indifferent of that needed to comprehend the Creator. We all are forced to question either the Creator and the Creation or our teachers.

If we for a moment pause and reflect on the consistency of the laws and principles of this creation and the consistency of those who have been the teachers of black people, we must ask: which one has been the most consistent and beneficial? Without a doubt, the creation has; therefore, it helps us to see that we must now rethink that which we have been taught, because our only education has come from our white slave-masters and their children.

"The American so-called Negroes are gravely deceived by their slave-master's teaching of God and the true religion of God. They do not know that they are deceived and do earnestly believe that they are taught right regardless of how evil the white race may be. Not knowing "self" or anyone else, they are a prey in the hands of the white race, the world's arch deceivers (the real devils in person). You are made to believe that you worship the true God, but you do not! God is unknown to you in that which the white race teaches you (a mystery God)."1

"The great arch deceivers (the white race) were taught by their father, Yakub, 6,000 years ago, how to teach that God is a spirit (spook) and not a man. In the grafting of his people (the white race), Mr. Yakub taught his people to contend with us over the reality of God by asking us of the whereabouts of that first One (God) who created the heavens and the earth, and that, Yakub said, we cannot do. Well, we all know that there was a God in the beginning that created all these things and do know that He does not exist today. But we know again that from that God, the person of God continued until today in His people, and today a Supreme One (God) has appeared among us with the same infinite wisdom to bring about a complete change."2

This belief in a God other than a man (a spirit) Allah has taught me, goes back into the millions of years - long before Yakub (the father of the devils) - because the knowledge of God was kept as a secret from the public. This is the first time that it has ever been revealed, and we, the poor rejected and despised people, are blessed to be the first of all the people of earth to receive this secret knowledge of God ...Any one so blind to the reality of God is the servant of the devil, until he or she sees God as a reality....3

"God means possessor of power and force."4 This means that those who are capable of exercising power and force over a things is in principle the god over it; yet, there is one that is more supreme. It's like one controlling one's own family, or one controlling many families, but the one who is controlling the universe is unquestionably the Supreme Being.

In this reality we live, and that is not to say there is another, if a thing exist, there is a way to measure it and this potential measurement dictates the accountability by the creator or maker of it; therefore, to ascribe a misunderstood process to supernatural control is always the height of error. No thinking person who respects his or her own intellect will fall back upon the occult excuse for an explanation of what he or she cannot comprehend.

The God of reality works through recognized forces, fixed in their laws and natural in their results. Superstition is merely the inability to account for causes. It is unreasoned awe, fear of something unknown, mysterious, imagery, or religious belief or practice founded upon fear or ignorance or a tenant, scruple or habit founded upon fear or ignorance.

Some religions prophesy a radical change in the future (the end of the world, the 1000 year reign of Christ, or a return to some golden age). One such "millenarian movement" is the cargo cult of Melanesia. The islands in Melanesia have been visited through history by foreigners in sailing ships, steamships, and airplanes. Each time the visitors brought riches unknown to the island people. The natives were greatly impressed by these goods, which they believed were being manufactured and sent to them by their ancestors.

The natives were most impressed with the cargo brought to the islands by planes during World War II (canned food, clothing, tools, radios, watches, guns, and motorcycles). When the War ended, however, the supplies were used up and not replaced by new shipments. The island prophets proclaimed that if airstrips were built, these would entice back the planes with their fabulous cargo. As a result, the natives have built airstrips in the jungle, complete with hangers, radio shacks, beacon towers, and airplanes (all made out of bamboo sticks and leaves). The airports are manned 24 hours a day, with bonfires set at night to serve as beacons for the expected cargo planes.

Thus, 40 years or so after World War II the natives await the phantom cargo that will hail the beginning of a new age. When the planes finally arrive, their lives will be filled with riches and plenty. They will be reunited with the deceased ancestors, heralding the beginning of heaven on earth.

I am pretty sure you are thinking of other religious doctrines of other religions, and some expressions closer to home, which fit this mode, and comparatively most of them believe in some force, spirit or thing outside of their comprehension.

The reality in which we all live dictates basic principles:

1. We all need basic elements to be born - sperm mixed with ovum.

2. We need food (water, nutrients, etc...) to sustain life.

3. Knowledge of how to live to preserve life.

Now, regardless of any prophet or messenger one believes in, the system from which they themselves came through and subsisted by was this established natural system we call reality. Therefore, if you think you have discovered a new system outside of this established reality, then the burden of proving it rest on you. The reality of God forces other realities.

CAN "REALITY" BE REINVENTED THROUGH TELEVISION?

"They" implanted the image, and "you" then carry it around inside you like some kind of neuronal billboard. There's nothing you can do about it if you're going to keep watching television at all. Your knowledge of real and non-real is useless. All images are "real images."

Think about this question for a moment; for regardless of the type of programs you may watch, if any, how much television do you watch per day? Some may watch only the news, nature programs or something "of importance;" however, how many generally take the time to add the time up?

Nevertheless, if people are ingesting television images at the rate of approximately four hours a day, then it is clear that whatever uses people have for the images they carry in their heads, television is now the source. It has become the most important single source of images in the world today.

When you are watching television, all categories of your own image-making capabilities go dormant, submerged in the television image. TV effectively intervenes between you and your personal images, substituting itself.

When watching TV, you are not daydreaming, or reading, or looking out the window at the world. You have opened your mind, and someone else's daydreams have entered. The images come from distant places you have never been, depict events you can never experience and possibly can't prove, and are sent by people you don't know and never met. Your mind is the screen for the microwave pictures. Once their images are inside you, they imprint upon your memory. They become yours. What's more frightening, the images remain in you permanently. Can this be proven you ask?

Please bring to mind any of the following: the Simpsons, Superman, John F. Kennedy, Batman, Bill Clinton or perhaps even Jesus. Did any of these images appear in your mind? Were you able to make a picture of them in your head? If so, that is proof that once they have entered your brain they remain in there. They live in there together with all the memories of your life. Yet you don't know most these people. And many of them are fictional characters.

Now, would you make the effort, please, to erase the TV people from your mind? Make them go away. Erase the Simpsons or Superman. Can you do that? If so, you are a most unusual person. Once television places an image inside your head, it is yours forever.

The Television image can be as real in effect as the personally experienced image. The mind doesn't automatically distinguish which image is from direct experience and which has been imposed by the media. If I should now ask you to erase the television image of the grocery store, leaving only the one you have experienced in reality - the personally experienced, direct contact - can you do that? Please make the effort and see if you can.

We are left with a very bizarre phenomenon. Television is capable of dominating personally derived imagery - from books or imagination - and it is also capable, at least some of the time, of causing confusion as to what is real experience and what is television experience. The mind is very democratic about its image bank, all are equally available for our recall and use. And so, when we call on our images for whatever purposes we may have for them, we are as likely to produce an implanted image as one which was originally our own.

The root of this unfortunate problem lies with the fact that until very recently, human beings had no need to make distinctions between artificial images of distant events and life directly lived.

Have you heard the axiom: Seeing is believing? Throughout the hundreds of thousands of generations of human experience, whatever we saw with our eyes was concrete and reliable. Experience was directly between us and the natural order or the natural environment, which is non-mediated, non-processed and not altered by other humans. If we saw a flock of birds flying southward, then these birds were definitely doing that. We could believe in it. We might interpret this concrete information in various ways, perhaps misinterpret it, but there could never have been a question as to whether it was happening. The information itself, the birds and their flight, could not be doubted.

This is the case with all sensory information. Whatever information the senses produce, the brain trusts as "inherently" believable. In other words, the eye ball itself don't think, it only transfers and processes the light which is sent through the optic nerve and the brain ascertains, identifies, defines and interprets the image. This belief in sense perception is the foundation, the basis, for human function.

The situation is clearer still when it comes to product advertising. When you are watching an advertisement, you know for sure that the advertiser is trying to get you to do something: buy the product. You also know that the people in the ad are not "real," that is, they are actors who are speaking lines, in situations that do not represent their actual lives. Everyone knows this. We all know that the motive of the sponsors, the actors and the writers of the ads are all trying to implant a feeling in us that will eventually get us to buy something.

We know that they are doing this but we very often act on the ad. Advertisers don't care at all if you know the advertising is fictional. They make very little effort to fool you about that, because whether or not you know it is fictional, the image of the product goes into your head. From then on, you've got the image and there's no letting it go.

If you then walk through a supermarket and spot the toothpaste that you've been carrying as an image, a little click goes off in your head. Familiarity. It doesn't mean you'll buy the toothpaste, but the "click of familiarity" goes off anyway. "They" implanted the image, and "you" then carry it around inside you like some kind of neuronal billboard. There's nothing you can do about it if you're going to keep watching television at all. Your knowledge of real and non-real is useless. All images are "real images."

In a sense, the advertising images are more real than other television images because you get to see the image "live" right in your supermarket. First you ingest the image of the toothpaste from television and file it. Then you see it in the store and you recognize it. (Have you heard your child say, “Hey, I saw that on television"? There's a real excitement being expressed.) If you buy the toothpaste, it's then right there in your bathroom, so the image from the screen materializes in your home. Advertisers are the alchemist of our day.

It works the same way, albeit more subtly, with the "behavioral" content of advertising and programs. You see Bill Cosby or the Waltons solve a family problem. You find yourself in a family situation which is not dissimilar. The image flashes past. You may reject it, but it flashes past nonetheless. If that's the only imagined instance you have available to call upon for such situation, you are somewhat more likely to be influenced by it. You don't interrupt your behavior to say, "Wait a minute; I've got to keep straight my bank of television imagery from my bank of real-world imagery." The mind doesn't work that way.

Now that the implanting of images is widely accepted as a norm, reshaping the viewer's reality is very simple. Do we not take for granted that when looking at a nature program the animals are where we are told there are. When viewing the animals in the jungle, how can you prove they exist unless you go there and witness the image yourself. It's like the child who asked if oranges and apples grew in stores. "Of course not," we tell them. "Fruit grows from the ground somewhere out in the countryside, and then it's put into trucks and brought to stores." But is this true? Have you seen that? When looking at the underwater programs, are there really fish down there. Have you actually seen the shark or the stingray.

What makes this so scary is that there is a large amount of experimental tomatocs, and the like, in grocery stores that have never come from the ground, but are manufactured in laboratories. There are movies depicting the impossible; however, through the use of technological breakthroughs in the field of making real look unreal, one can produce the image of an entire city underwater (Sea Quest) or entire cities in the sky (Deep Space 9 and Babylon 5) equipped with "other life forms."

If the future shock of these movies are aimed at producing a hope that these things are possible in the future, and the world as we have come to know it has a future, then is it safe to assume that people will copy what they see?

The people are now accepting on face value that many images shown to them about things far away are real and believable. How many of us went to see what was really happening in Vietnam? How many of us went to the Moon to make sure it wasn't a mega-trick staged by the same producers of Star Trek? How many of us went deep-sea scuba diving to make sure the whales we saw on television actually were where National Geographic said they were? How many of us can drill underground to see if the devil and hell are under there as this world have been telling us for the past 6,000 years and lastly, when do you intend to take a ship into the sky to make sure the heaven, which the preacher told you and me of is really up there? Or have you settled for the image shown on TV? And if this is not your image of this, what image do you have and how can you prove it exists to yourself or to anyone else?

Do you sometimes feel that everything you do is ineffective? Or, that you seem to be running on a treadmill; whereas, as you advance, you are always in the same place? Or, that every issue had to be fought as though it were the first one. People seemed unable to connect one issue to another, to find common threads in, say, a struggle against anything. Not even specific victories were possible, and overall understanding of the forces that were moving society seemed to be diminishing.

People's minds seem to be running in dogged, one-dimensional channels that are scary when one begins to see that this is an externally motivated attitude projected to and absorbed by all who wish to feast. Modern technology is all but relieving the people of self-dependence on a daily basis with the society as a whole moving more and more to a service oriented set up; in other words, someone else is attempting to do it for you and make the price of "doing it yourself" harder and harder to afford. It is becoming easier and easier not to leave home. Could life within these new forms of physical confinement produce mental confinement?

Television is the main tool promoting and shaping the whole scenario. This tool allow what's unreal to appear real and what's real appear unreal. In the nature of television imagery, it allows form to supersede content and advertising is the field that pioneered conveying "assumed content" in place of substance. The mistake most black people make is trying to use it to deliver "content" through a medium predisposed to resist it. Many black leaders today have become deceived into believing that in order for them to get to our people, they must be on television or be "piped in" on someone's program or talk show. What most of them fail to realize is that they themselves become media illusions. Once this is in place, their gravest mistake would be to assault the machine. The machine doesn't care about its fantasies. A new one will always do. The only goals of the machine are to continue to be the real power behind the throne, no matter who is king, and to remain the primary factor in all public perception.

There has been many events later exposed as only one of the many non-events pushed through the media to sell the public. The fact that this could be done at all - fictional news about fictional events, military or otherwise, expanding faraway things that no one watching the images could observe firsthand - is cause for serious alarm about the power of the media to pursue fictitious realities.

To even pursue being in the media for any length of time, it requires a mind change. What many of our people don't know is that in order to take a fight or struggle up on a white media, it requires a white mind; moreover, to fight the white man on his media means a simultaneous process of self-destroying what remains of their blackness - they become watered down and gray, and very vague about principles and positions they once were very clear and decisive about.

The white man didn't create the media machine to disseminate reality or truth; therefore, when one attempts to convey something subtle, complex, foreign and ancient through a medium, which doesn't seem able to handle any of that and which is better suited to objective data, conflict and fast packaged information, their efforts are doomed to failure.

There is too much information available at once and that the people are being given too much at one time. The population is being inundated with conflicting versions of increasingly complex events. People are giving up on understanding anything and relying more and more on sound bites and accepting them as truth and reality. Could it be that this glut of information is dulling awareness, not aiding it. Overload. It encouraged passivity, not involvement.

We mentioned an analogy before concerning a question by a child asking whether apples and oranges grow in stores. "Of course not," we tell them. "Fruit grows from the ground somewhere out in the countryside, and then it's put into trucks and brought to the stores."

But is this true today in these times of genetic engineering? Have you seen that? Do you have a sense that what you are eating was once alive, growing on its own?

We learn in school that fruit grows from the ground. We see pictures of fruit growing. But when we live in cities, confined to the walls and floors of our concrete environments, we don't actually see the slow process of a blossom appearing on a tree, then becoming a bud that grows into a apple. We learn this, but can't really "know" what it means, or that a whole cycle is operating: sky to ground to root through tree to bud ripening into fruit that we can eat. Nor do we see the particular value of this knowledge. It remains an idea to us, an abstraction that is difficult to integrate into our consciousness without direct experience of the process. Therefore we don't develop a feeling about it, a caring. In the end, how can our children or we ourselves really grasp that fruit growing from trees has anything to do with humans growing from eating fruit?

While many attempt to teach about a natural or organic environment out there somewhere, the artificial environment that we are in is teaching us that nature is irrelevant, separate from us, and of only intellectual value. The natural environment, if it existed for us anywhere, was only in our minds, in our memories.

The Honorable Elijah Muhammad teaches us that submitting to the natural laws of the Creator is the Nature in which we, black people, are created. Submitting is called Islam in the Arabic language. What this means is that the basis on which television rest is against our very nature. It's dangers vary. It enters inside human beings, inside our homes, our minds, our bodies, making possible the reordering of human processes from the inside.

It substitutes for diversity of experiences and once it is reduced to television, a relative handful of people can control one's awareness. Whereas, whoever is in a position to pay for the technology has primary access to your awareness. This media machine is allowed to pass for experience. It is anti-experience. Its interaction with the human body and mind fixes people to itself, dulls human sensibility and dims awareness of the world. By focusing people on events well outside their lives, television encourages passivity and inaction, discourages self-awareness and the ability to cope personally.

Television encourages separation: people from community, people from each other, people from themselves, creating more buying units and discouraging organized opposition to the system. It creates a surrogate community: itself. It becomes everyone's intimate friend, advisor, teacher and guide to appropriate behavior and awareness. Thereby, it becomes its own feedback system, furthering its own growth and accelerating the transformation of everything and everyone into artificial form.

Our minds are being thoroughly striped from any desire to have it taxed. Digital Video Discs (DVD's) and Compact Discs (CD's) have become the study tools of choice. Instead of mentally stimulating interchange with a book, writing and oral discussion, we have come to prefer to hear it on disc, radio or television and accept the information as truth on face value. When we encounter one another, more fear is produce in us by one finding out how weak we are when it comes to successfully defending what we say we believe in, with analytical facts, intellectual interchange and literary prowess (explicit knowledge of the Messenger's teachings, Holy Qur'an, or Bible, etc...).

We subscribe to the "assumed content" or "official lines" we received from the Sunday rostrum, latest CD or DVD. This great need to “stay up" on the latest is what keeps a major part of the revenues coming in. Substance is sacrificed on the alter of cosmetics, glitter, glow and oratorical showmanship. The greatest testimony to this is how deficient the adherers are. They are spiritual skeletons or as the Bible calls them, “Wood clad with garments.” All noise by no substance.

We must get back to the pursuit of substance and content and divest ourselves of the degenerating effect of television and it's promoters. We become what we see. And in today's America, what most of us see is one heck of a lot of television. If television passes for experience and provides us with a false sense of reality, which in essences shapes our thinking, have you not heard that as a man (or woman) thinketh so is he?

Thank you for purchasing this book. We trust the reading was rewarding and enlightening.

We offer various titles by Minister Nasir Hakim, as well a comprehensive collection of Messenger Elijah Muhammad's works. These works include, but are not limited to:

- **Standard Published Titles**
- **Unpublished & Diligently Transcribed Compilations**
- **Audio Cassettes**
- **Video Cassettes**
- **Audio CD's**
- **DVD's**
- **Rare Articles**
- **Year Books**
- **Annual Brochures**

You are welcomed to sample a catalog of these items by simply requesting a FREE archive Catalog.

Our contact information is as follows:

Secretarius MEMPS Ministries
12685 Dorsett Rd. #187
Maryland Heights, Missouri 63043
Ph./Fax 314.564.4003 ● secmm@sbcglobal.net

Wholesale options are also available.

Made in the USA
Monee, IL
15 September 2026

63208996R00046